Can Mud Be THAT?!

Sandra Markle

Illustrated by

Alexandra Colombo

Charlesbridge

Think you know MUD?

It's squishy, gooey, and sometimes yucky, yes? *But WAIT!* To many different animals, mud can be so much MORE!

Can a mud puddle be . . .

a slushie? Yummy!

Swallowtail butterflies visit mud puddles for a healthy drink.

A butterfly sips with its long, straw-like mouthpart. But it needs more than a drink of water. To be healthy, the butterfly needs the salts and minerals from the dirt that made the puddle muddy.

But WAIT! There's more!

Can a mud puddle also be . . .

face paint? *Look at me!*

An Egyptian vulture dips its head in a mud puddle for a splash of color.

Mud turns the vulture's white head, neck, and chest feathers red. No one knows for sure why it paints itself. Maybe to become a pretty bird—and attract a mate!

But WAIT!

Can sloppy mud also be . . .

sunscreen? *Slap it on!*

An African elephant takes mud baths to protect its skin.

An elephant's skin is almost an inch thick in most places! And it has very little hair to shield it from getting a bad sunburn. So the elephant rolls in mud and sprays it on, too. Sloppy mud is nature's perfect sunscreen.

But WAIT!

Can sloppy mud also be . . .

a cool pool? *Please don't pass gas!*

A hippo soaks in sloppy mud to keep from overheating.

A hippo cannot sweat the same way humans do. So it escapes its hot habitat by hopping in a mud bath with its herd. Come evening the hippo exits the mud to graze on grasses. It has a big appetite, which can mean lots of passing gas and pooping—sometimes while soaking!

But WAIT!

Sometimes mud is a little sludgy! Can this kind of mucky mud be . . .

a shelter? *Ooze on in!*

A snapping turtle digs into a pond's mucky mud to escape harsh winter conditions.

Once settled, its body slows down enough to need very little oxygen. The turtle skips eating while food is scarce. Come springtime it gets back to swimming, surfacing to breathe and gobble up whatever it catches.

But WAIT!

Oozy mud can be a little gooey, too. Can gooey mud be . . .

bug repellent? *Shoo, fly!*

A warthog rolls around in mud for protection.

Wet mud cools a warthog on a hot day. Wet or dry, a coat of gooey mud creates a shield on the warthog's skin. This keeps pesky bugs from biting.

But WAIT!

Can gooey mud also be . . .

glue? *Stick tight!*

A beaver grabs mud from the pond bottom with its front paws to build.

A beaver pair uses mud to pack sticks together to build a dam across a stream. The dam blocks water flow to create a pond. It's the perfect place for the beaver pair to build their lodge home—with more mud and sticks—before raising their kits!

But WAIT!

What if mud is even thicker and stickier? Can sticky mud be . . .

a nursery? ***Welcome, babies!***

A female mud dauber wasp uses mud to build a safe place for her eggs.

The wasp scrapes up a dab of sticky mud with her toothlike mouthparts. She flies this to a sheltered spot and, little by little, shapes the mud into tubes. Inside these tubes she packs food—insects and spiders—and lays her eggs. This is where her young will hatch and grow up, eating the food their mother left them.

But WAIT!

Can sticky mud also be . . .

a highway? *Roadwork ahead!*

Subterranean termites create mud tubes for safe travels.

These termite colonies live in underground nests, but they must find food above ground. So the termites mix dirt with their saliva and waste to make sticky mud just right for building tubes. When the moist mud hardens, the termites can move through the tubes, from their nest into the world above.

But *WAIT!*

What if sticky mud is less watery?
Can dense, soggy mud be . . .

a castle? *Home of Chick Charming!*

American flamingo parents build mud nests atop mud to keep their one chick safe.

Flamingos nest in thick, soggy areas where water meets land. Sometimes these mudflats can flood. To keep their single chick safe, flamingo parents pull mud with their big, crooked bills until they build a mound as high as twenty-four inches! Their chick stays dry until it learns to swim.

But WAIT!

Can a mudflat also be . . .

a restaurant? *Self-serve only!*

Lots of hungry birds visit mudflats to eat their fill.

Some have long beaks. Others have short. Some beaks are curved. Others are straight. But each bird's beak is just right for pulling a crab, shrimp, or worm meal from the mudflat.

But WAIT!

What about you?

Can you use mud for . . .

Dishes? Art? Construction? And skin care? *You bet!*

Humans use mud to mold pottery for dishes and cookie jars. They make jewelry, garden gnomes, and bases for lamps. They also use mud to create bricks for buildings, and to soak and soothe their bodies at the spa!

For animals and humans alike, mud has many uses.

But maybe the BEST way for you to use mud is for . . .

PLAY!

MUD-tastic Facts!

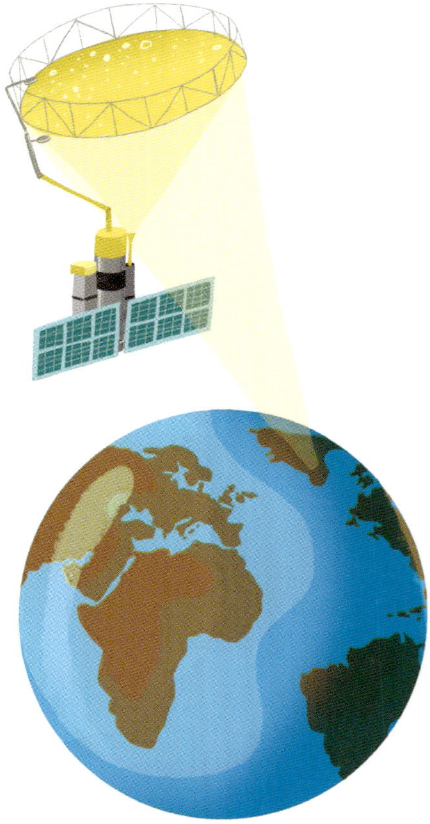

- Mud keeps the Taj Mahal in India white. Every few years, workers carefully spread a special mud paste on the marble. Then they wash it off. This process removes a yellow stain on the white marble caused by air pollution.

- In 2015, NASA put a satellite into space to map Earth's muddy places. The goal was to better understand where there could be long dry spells and floods.

- Starting in the 1930s, the balls used for all Major League Baseball games have been rubbed with mud so they are not slippery. It is a special smooth mud found along just one spot on the Delaware River's banks. The exact spot is a company secret.

- Azerbaijan has more than 400 mud volcanoes—more than any other country. A mud volcano forms when a pocket of gas builds up underground until the pressure forces the mud to ooze or blow out between fractures in Earth's surface.

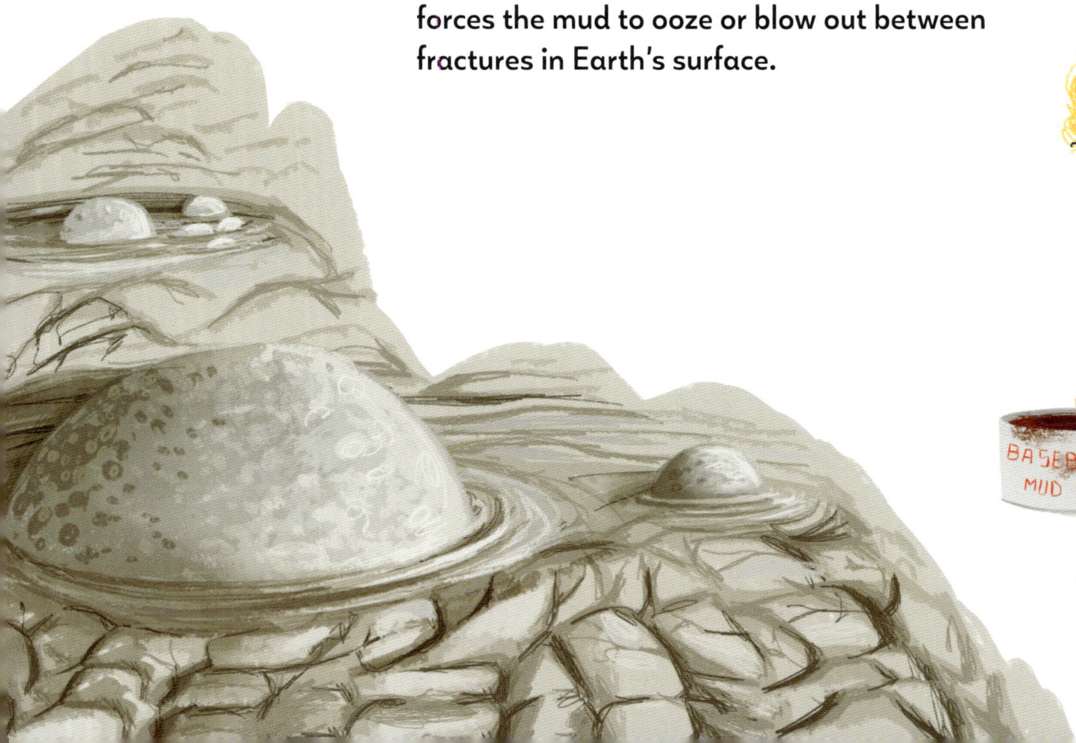

Playtime IN the MUD!

This is messy FUN! Work with an adult partner to choose an outdoor spot for muddy play. Next add a plastic kiddie pool, wheelbarrow, or big bucket for mud mixing. Don't dig up dirt. Use a bag of lawn-and-garden topsoil that is sure to mix well and be free of rocks, bugs, and earthworms. Fill empty, clean milk jugs with water for mud-making. Then collect a wooden spoon or stick for stirring. Now dress to be messy in washable, old clothes and boots or old shoes. And start playing!

CASTLE BUILDER: Dump the bag of soil into your mix site. Add water slowly as you stir to create thick MUDFLAT MUD. Shape this with your hands to build your own mud castle. Make it fancier than a flamingo's castle nest. Add a tower, balconies, and more!

MUD BAKER: Use more topsoil or return your first MUDFLAT MUD to the mixer. Add more water slowly, and stir until you have STICKY MUD or GOOEY MUD. Now shape mud pies, mud cookies, mud cakes, mud donuts, or anything else you like. Try making a dozen. BUT DON'T EAT ANY!

MUD ARTIST: Return your bakery goodies to the mixer, or get more topsoil. Add more water slowly, and stir until you have SLOPPY MUD or MUD-PUDDLE MUD. You'll also need sheets of white construction paper. Place the paper on the ground. Dip just your fingers or press your hand palm side down into the mud you made. It's now paint. Draw. Splatter. Make handprints. Create mud art! Let these dry to keep and display!

For Alana Stanton and all the children at Oconee County Primary School in Watkinsville, Georgia.

A special thank you to Skip Jeffery for his loving support during the creative process.—S. M.

For Alessia: We have been friends since we were five years old, and we still know how to have fun in the mud.—A. C.

Charlesbridge • 9 Galen Street, Watertown, MA 02472
www.charlesbridge.com

Printed in China • OPIC
The authorized representative in the EU for product safety and compliance is eucomply OÜ, Pärnu mnt 139b-14, 11317 Tallinn, Estonia, hello@eucompliancepartner.com, +33757690241
(hc) 10 9 8 7 6 5 4 3 2 1

Illustrations done in Photoshop
Text type set in Blauth by Daniel Hernández Luciano Vergara
Edited by Julie Bliven
Designed by Cathleen Schaad
Production supervised by Jennifer Most Delaney

Library of Congress Cataloging-in-Publication Data
Names: Markle, Sandra author | Colombo, Alexandra illustrator
Title: Can mud be that?! / Sandra Markle; Illustrated by Alexandra Colombo.
Description: Watertown: Charlesbridge, [2026] | Audience: Ages 3–7 | Audience: Grades K–1 | Summary: "A silly, splashy—and absolutely accurate—ode to all the unexpected things mud can be for a variety of different animals, including food, shelter, protection, and more." —Provided by publisher.
Identifiers: LCCN 2025029571 (print) | LCCN 2025029572 (ebook) | ISBN 9781623546380 hardcover | ISBN 9781632892850 ebook
Subjects: LCSH: Mud—Juvenile literature | Animals—Juvenile literature
Classification: LCC QE471.15.M83 M37 2026 (print) | LCC QE471.15.M83 (ebook)
LC record available at https://lccn.loc.gov/2025029571
LC ebook record available at https://lccn.loc.gov/2025029572